The Animal Kingdom
ANIMAL HOMES

Malcolm Penny

Illustrated by Robert Morton

The Bookwright Press
New York · 1987

The Animal Kingdom

Animal Evolution
Animal Homes
Animal Migration
Animals and their Young

First published in the
United States in 1987 by
The Bookwright Press
387 Park Avenue South
New York, NY 10016

ISBN 0–531–18122–7
Library of Congress Catalog Card Number: 86–73113

First published in 1987 by
Wayland (Publishers) Limited
61 Western Road, Hove
East Sussex BN3 1JD, England

© Copyright 1987 Wayland (Publishers) Limited

Typeset by DP Press, Sevenoaks, Kent, England
Printed by Casterman, S.A., Belgium

Contents

What is a home?	4
Holes in the ground	6
Homes of mud	8
Palaces of mud	10
Homes of sand	12
Burrows and warrens	14
Homes of grass	16
Homes of sticks	18
Homes of leaves	20
Homes of wood	22
Homes by water	24
Animal squatters	26
Living with humans	28
Glossary	30
Further information	30
Index	32

What is a home?

Below *Kingfishers nest in holes in banks, while rabbits live in warrens. Spiders produce their own silk for making webs. Butterflies, however, have no home, but shelter under a leaf at night.*

When humans build a home, they use the materials that are easily available. In cities, artificial materials like glass and concrete are used. In very remote places, however, like the African bush or the Arctic, homes are built from natural materials that can be found nearby. In the bush the Africans use sticks, mud and grass, while in the Arctic the Inuit use snow blocks. Both make excellent houses.

Animals that need homes make use of materials from their surroundings. However, most animals use their homes very differently from the way we humans use ours. Some animals need homes for only part of the year, usually during the breeding season. Birds are a good example, making nests in the spring and summer and then being homeless for the rest of the year.

Other animals live in permanent homes throughout the year, like rabbits in a warren or ants in their nests. There are animals that have no homes at all – antelope on the African plains and whales in the oceans have no special place to which they regularly return, to rest or to rear families.

There is a fish called a blenny that is often found taking shelter in bottles that have been thrown into the sea. We cannot say that it has built its home, nor have its neighbors the hermit crabs, which walk about on the sea floor wearing the shells of dead whelks for protection.

All the animals in this book have made their home by digging or building, weaving grass or piling up sticks, or by using other materials in interesting ways. Some animals live alone, others share a home.

This black goby fish has used an empty can for its home.

Holes in the ground

The simplest kind of home that an animal can make is a hole in the ground. Voles and mice are not specially built for digging, but they make holes in which to shelter and to bring up their young. Birds like puffins and sand martins make their nests in the same burrows year after year.

Many wasps and bees live alone, not in crowded colonies. They often dig holes in soft earth in which they lay their eggs. Many dung beetles also do this. The tiger beetle is different – the female lays her egg in the soil, and the larva digs a burrow for itself when it hatches.

Bears hibernate during the winter in their dens. They emerge in spring, when their cubs are born.

Wombats are very sturdy creatures with powerful feet for digging out their long burrows.

Really expert diggers have specially adapted feet and powerful legs, like the mole, which has front feet with long curved claws shaped like shovels. It also has tiny eyes and thick velvety fur. These are ideal adaptations for a creature that spends most of its life underground.

Some large mammals make huge holes. Bears dig underground dens, often in the side of a hill. Inside they are more than three feet high. Polar bears do the same in snow banks to shelter from the freezing Arctic winter. The wombat in Australia makes a burrow as much as 5m (16ft) long. Badgers are smaller, and so make smaller holes, but they have similar shovel-shaped feet and powerful short legs.

Surprisingly, some of the biggest underground homes are made by very small creatures. Large colonies of leaf-cutter ants in South America dig out underground cities that may be 20m (66ft) across. To make this vast hole, the ants remove the soil one grain at a time.

Homes of mud

In hot countries, the inside walls of buildings are often dotted with neat balls of dried mud, sometimes joined together in rows. They are the nests of the potter wasp.

The wasp begins by collecting pellets of wet mud from the edge of a puddle or stream. It rolls the pellets between its head and its legs. Then it flies with the mud pellets to the chosen nest site in a building or on a tree. There the wasp molds the pellet with its jaws into a ribbon of mud, which it sticks to the wall or tree.

At first the nest is cup-shaped, but as the work goes on, the open end becomes narrower until it forms a neck, finished off with a neat rim. Before closing the neck with a final pellet of mud, the potter

A pair of ovenbirds have just finished building their dome-shaped mud nest.

wasp lays a single egg inside, together with several paralyzed caterpillars. When the egg hatches, the grub will eat the caterpillars. Later it will develop into an adult wasp and break out of the pot.

An interesting mud-building bird is the ovenbird of Central and South America. A pair of adult birds gathers lumps of mud which they mix with straw and cow dung. Using their beaks and feet, they build their dome-shaped nest with the mixture, leaving a small entrance hole in one side. The mud nest is baked hard in the hot climate.

Swallows and house martins are other birds that build nests of mud. They gather it in their beaks from muddy places beside pools or streams. Swallows usually build their shallow cup nests inside a farm building, and so in North America they are called "barn swallows." House martins build their nests on the outside walls of houses, sheltered under the eaves.

Potter wasps gathering mud for their nests. Their beautiful vase-shaped nests have been copied by American Indian potters.

Palaces of mud

The best mud builders are termites, insects about the same size as ants. Many termites live underground or in trees, but some species in Africa and Australia build huge mud palaces, up to 7m (23 ft) high.

The colony starts as a tiny hole in the ground, where a newly mated queen termite begins to lay her eggs. The eggs hatch into worker termites, which will build the palace while the queen goes on laying eggs, as many as thirty thousand every day.

At first the building remains underground. The workers dig tunnels to connect the chambers where they look after the new babies. Some tunnels lead away from the colony. The workers use them as roads when they go out to collect food.

In the South American rain forest, a tamandua anteater feeds on tree termite nests. On the ground are pagoda-shaped termite mounds.

After a year or more, when there are several hundred thousand workers in the colony, they begin to build above ground. They gather fine soil and mix it with their own droppings and saliva into balls. The mixture sets harder than concrete, as the workers pile up the balls like tiny bricks. Soon there is a network of tunnels and chambers, enclosed in a thick and very hard outer wall.

As the castle grows and the number of workers increases to over a million, it becomes hotter inside, not just from the sun, but from the activity of the termites. So they build ventilation shafts and cooling chambers, sometimes with tall chimneys, to release the hot, stale air.

The final size and shape of the termite mound depends on the species that builds it, but they are all masterpieces of insect architecture.

Above *A close-up view of termites at work in their nest.*

In hot countries termites build enormous mounds, like this one in Queensland, Australia.

Homes of sand

Above *A sand wasp seals the burrow with her egg and a caterpillar inside.*

Sand is a very difficult material for a builder to use because it tends to collapse unless it is stuck together in some way. Different animals solve this problem in different ways.

The sand-mason worm glues grains of sand together with a sticky substance that it produces from its skin. Reaching out, the worm covers its tentacles with sand and then withdraws them. This leaves the grains of sand stuck to one another in a tube. If you walk on the beach at low tide, you can see the sand-mason worm's tubes sticking out of the sand like little flowers. The worm lives under the sand when the tide is out.

Above *At night a trapdoor spider waits in its hole (left) and catches a passing beetle (right).*

Left *A fiddler crab in an Indonesian swamp plugs up its hole to keep out the water as the tide comes in.*

Fiddler crabs of tropical swamps and coasts make burrows in sand or thick mud. At low tide the crabs live on the surface of the mud, but at high tide they go into their burrows. To keep out the water, they plug up their holes with a ball of mud.

Some wasps burrow in sand. They usually choose a place where the sand is hard and already stuck together, such as the bottom of a dried-up lake. The female closes the burrow with a tiny pebble, to protect her egg and the supply of caterpillars for the wasp grub to eat.

The trapdoor spider is a very clever sand builder. It lines its burrow with sticky saliva covered with the silk the spider produces. Then it mixes plenty of soil particles and silk to make a hinged lid about $1/3$ inch thick, to seal the burrow like a bath plug. The spider never leaves the burrow to catch its prey. Instead it waits until it senses that a passing insect is close to its hole. Quickly the spider pushes up the lid, grabs the insect and retreats inside again.

Burrows and warrens

A rabbit warren is much more than just a hole in the ground. It is a system of linked burrows with many entrances. Most of the digging is done by the females when they are preparing nests for their litters of babies. The mother lines the nest with fresh grass and with her own fur. She shuts the babies in with a wall of earth and visits them regularly to feed them. The babies break out of their nest when they are about three weeks old, often making a new exit for the warren as they do so. In the autumn all the rabbits dig to extend the warren, so that it can shelter the growing colony through the winter.

A cut-through view of a marmot warren, showing the separate chambers for babies and droppings.

The platypus uses a hidden underwater entrance to reach its eggs, which are kept above the water level in the long burrow.

A badger family lives in a complex of underground passages, called a set. The badgers are good at digging and also work hard to keep the set clean and airy. They make small air holes in the tunnels and often change the grass linings of the chambers.

The marmot, a rodent found in the mountains of North America, Asia and Europe, is a great warren builder. Marmot warrens extend about 3 m (10 ft) deep, with many linking passages and separate chambers for droppings. Marmots line their sleeping chambers with soft grass. In winter the outer part of the warren is packed with dried grass, earth and stones. This keeps out the cold while the marmots hibernate until the spring.

The duckbilled platypus in Australia digs a long horizontal burrow with a nesting chamber at the end, where it lays its eggs. The burrow leads away from the pool or stream where the platypus feeds, and has a hidden underwater entrance.

15

Homes of grass

Below *The nests of the harvest mouse and reed warbler blend in well among the leaves.*

Many small birds, such as warblers, build cup-shaped nests from grass. The basic method of building with grass is the same all over the world.

The bird wedges the first few pieces of thick grass into a suitable place until they make a platform. It then collects finer grass blades and piles them onto the platform. Next it turns around and around in the middle of the pile to hollow out a bowl, before pushing the sides into position with its feet. All the time it keeps turning around, so that the bowl remains circular.

The next step is to pull out the ends of some of the grass blades one at a time, and push them back with its beak into a different part of the side of the nest. This weaving makes a strong, basket-like nest.

The finishing touch for many birds is to bind the edges with strands of sticky spider's web.

Weaverbirds, a kind of finch from Africa, make much more complicated nests. They begin by making a ring of grass hanging from a twig on a tree. Then they add more grass, closely woven together, until the whole nest is shaped like a hollow ball, with an entrance in the side. The entrance is usually protected by a porch. Often one tree is covered with many weaverbird nests.

The strongest grass nest of all is made by the harvest mouse. The mouse splits green wheat leaves lengthwise, and pulls loops from one leaf through slits in another leaf. In this way it knits a ball of living leaves attached to wheat stalks. The nest is perfectly hidden because the leaves of the nest are still growing. So the nest stays green until the wheat ripens, and then it turns brown.

Above *A weaverbird works on its nest in a thorn acacia tree in Africa.*

Homes of sticks

A pair of ospreys have nested on a bare treetop by a lake. Nearby in a pine wood, a red squirrel has built its dray. A jackdaw (bottom right) has also nested here.

Most larger birds build nests of sticks, instead of grass. Jackdaws and other members of the crow family, such as magpies, collect twigs which they pile into the fork of a tree. The platform that they make is surprisingly firm. This is because the birds shake the twigs as they put them in place, causing the small side-shoots to become tangled and interlocked. Really big birds, such as storks, and especially eagles, use very large sticks to build their nests, which are called aeries.

Eagles often use the same nest, or aerie, year after year. They add more sticks each time they return until the pile may be 2m (6ft) thick. The biggest stick nest of all is made by the hammer-kop, or hammer-headed stork, in central Africa. Strengthened with

mud and tufts of grass, it is 1m (3.3ft) across and nearly 2m (6½ft) tall. The chamber inside, at the end of a neatly plastered entrance tunnel, is 30 cm (12 in) high and 60 cm (2ft) across. The whole nest weighs 100 kg (220 lb).

Not all stick nests are built by birds. Squirrels build nests called drays high in trees. They do not use the nests only in the breeding season as places to rear their young. During the rest of the year the dray is used as a place to sleep, where the squirrels will be safe from owls and other predators.

Although they are not strictly homes, it would be a pity to leave out the sleeping nests of chimpanzees and orangutans. They usually build a new one each night, bending and weaving springy branches into a hammock where they can sleep in great comfort. Gorillas arrange leafy branches into a nest on the ground as darkness falls.

Homes of leaves

Tailor ants in Australia work as a team to build their nests, by pulling the edges of living leaves together. Other ants inside the nest stick the edges together, using their silk-producing larvae like tubes of glue. The adults carry the larvae to the nest site and attach the silk to one leaf and then the other alternately. The silk shrinks as it dries, forming a tight seam between the leaves. So the ants make themselves a well-hidden, waterproof home.

Shrinking silk is used by the remarkable leaf-curling spider, also in Australia. It makes its home inside a twisted, dead eucalyptus leaf, spinning a web that connects the opposite edges of the leaf.

Some tailor ants pull leaf edges together, while others use the silk from the larvae as glue to join the leaves.

Above *A tailorbird sews two leaves together with a strand of spider's web (left). The fleecy lining of the nest keeps the chicks warm (right).*

As the silk dries, it pulls the edges together to form a spiral tube. To make its home even safer, the spider picks up the leaf from the floor of the forest, and hangs it from a rope of silk before it begins to curl the leaf up.

The tailorbird, from India, China and Southeast Asia, is another animal that builds its home from living leaves. It has a very sharp beak, which it uses to make holes in the edges of two leaves. It then threads spider's web, or strips of tree bark, through the holes, like a shoelace through the eyelets of a shoe, until the two leaves are sewn together to form a bag. Finally the bird fills the bag with sheep's wool or dry moss, to make a snug nest for its young.

Sometimes a tailorbird chooses just one large leaf for its nest. It folds and twists the leaf into a bag shape and then sews up the edges.

Homes of wood

Although a number of different animals live in holes in trees, only two sorts, both birds, can actually make their own holes. They are woodpeckers, which live all over the world, and their relatives, the barbets, which live in tropical rain forests.

Both woodpeckers and barbets chip away at dead trees to find their food, mainly burrowing insects. The strong sharp bills, which the birds need for feeding, enable them to dig out more wood, to make a hole big enough to live in. The woodpecker's bill works like a chisel, chipping off flakes of wood, as the bird hollows out a hole.

A woodpecker's nest has an entrance just big enough for the adult bird to slip through. Inside, the hollow nest becomes wider as it gets deeper.

Below *The spotted-crowned barbet (center) and the Guatemalan woodpecker (right) make their holes. The keel-billed toucan (left) lives in empty holes.*

Above *Some wasps make their papery nests in human homes, for example outside under a roof or inside a loft. A finished nest can be quite large, as shown in the picture.*

At the bottom there is room for the bird to sit on the eggs. The base is covered with wood chips to provide a soft, warm cushion for the eggs, and for the chicks when they hatch. Woodpeckers are unusual birds because most of them use their nests not only for rearing young, but for roosting in at night, all year round.

Another group of animals that need wood to make a home are certain kinds of wasps. They chew the wood into fine fibers, mixing it with their saliva until it becomes a smooth paste. They carry balls of this paste to the nest site, where they mold it with their jaws to form the thin walls of their six-sided cells. One cell is joined to another, so a large nest can be made. When the paste dries, it forms a light stiff material which is very familiar to humans – we call it paper. Sometimes wasps nests like these are found in houses.

Homes by water

A water spider inside its air-filled diving bell.

Fish live underwater, but most of them do not build homes. They usually find shelter among rocks or in coral reefs, though some of them dig burrows in the seabed. A particular kind of goby fish makes its home by digging a trench, usually under the edge of a rock. It builds a wall in front of the trench, by collecting pebbles in its mouth and piling them up.

The water spider builds an underwater home. It spins a web between the stems of waterweed and fills it with bubbles of air. The spider brings down air from the surface which is trapped by the hairs on the spider's abdomen, forming a bubble. The water spider uses its home like a diving bell, making hunting trips from it and bringing back its prey to eat there. Later, the female lays her eggs in the air-filled diving bell.

Beavers build their homes, called lodges, right in the middle of a stream, out of reach of their enemies. A beaver lodge is a pile of sticks with two or more underwater entrances and a living space inside, above the water level. Before they build their lodge, a colony of beavers gnaws away at tall trees until they are felled. They then build a log dam across the stream, to slow down the fast-flowing water, turning it into a small lake. The still waters of the lake will not sweep the lodge away. Instead they surround it, like a moat around a castle.

Most water birds build their nests close to water, but grebes build their nests on the water. Their floating platform nests are anchored to water plants. Because they float on the surface, these nests are not flooded when the water level rises.

Right *Some beavers building their lodge and dam. Among the reeds a grebe sits on her floating nest.*

Animal squatters

This starling has taken over an empty hole made by a woodpecker.

When an animal builds a home for itself, it is also providing a home for other creatures. Every animal shares its home with parasites, such as fleas and lice, which live on the animal's blood. Unlike parasites, however, many animals are more interested in the home than its builder. We could call these animals squatters, because they are living on someone else's property.

The enormous roofed stick nest of the hammer-headed stork shelters several other animals, when the bird has finished with it. Egyptian geese nest on the roof and acacia rats cram thorny twigs into the entrance passage, to keep out predators. Sometimes a gray kestrel moves in before them.

Termite mounds provide a home for mongooses, which live in the ventilation shafts, and for a tiny snake which glides through the passages, feeding upon the termites.

Foxes spend the winter in an underground home called an earth. However, the fox rarely digs out its earth. It usually takes over a disused rabbit warren or even a badger's set.

Because woodpeckers are so talented at making holes in trees, their old nests are used by many other creatures, from birds like parrots and quetzals, to wasps. Sometimes starlings will wait until the woodpecker has hollowed out its hole and then move in before the poor woodpecker can use it.

The most famous intruders of all are cuckoos, which never build their own nests. They lay their eggs in the nests of other birds and leave their chicks to be fed and reared by the birds that built the nest.

Right *A pair of African gray kestrels use an empty hammerkop's nest, while below them some mongooses invade a termite mound.*

Living with humans

Which animals live in homes with glass windows, electric lighting and a running water supply? The answer is not just humans. In every human home there live many other animals – some are visitors, others live there all the time; some are harmless, while others are a real nuisance.

Woodworm and deathwatch beetles, which can also live in trees, sometimes live in our homes. They bore holes inside wooden beams and furniture to make their homes.

Some creatures depend on humans for food and shelter. The house spider, the house mouse, the carpet beetle and the clothes moth would find it hard to live without us. Although it makes messy webs and scares some people, the house spider is useful to us because it catches pests like flies. The bright-eyed house mouse, on the other hand, breeds rapidly, gnaws away at floor boards and can cause unpleasant smells.

Carpet beetle larvae are also destructive, making threadbare patches in blankets and carpets.

Below *Some visitors to human homes. Flies bring disease, but may be caught by house spiders. Mice make holes in boards and raid our food stores. Woodworm beetles bore holes in wooden furniture, while carpet beetles eat holes in carpets.*

In the wild, house martins nest on cliff faces, but in built-up areas they nest on the walls of houses under the eaves.

Clothes moth caterpillars make holes in sweaters and blankets. The moth lays its eggs in woolens because wool provides a home and food for the caterpillars, which hatch in the spring or summer.

In some homes, bats may be found living inside attic rooms, or house martins may nest on the outside walls under the eaves.

As we have seen, animals can make their homes almost anywhere. Those that live in our homes make rather simple dwellings out of our materials. Those that live in the wild, however, are often more skillful. Some, like the weaverbird and the beaver, are master craftsworkers.

Glossary

Adapted Changed to suit a particular purpose.
Aerie The nest of an eagle or other bird of prey. Aeries are built on high treetops or on rocky ledges.
Architecture A building or structure.
Arctic The very cold northern region of the earth that lies within the Arctic Circle.
Bush A name given to the hot, shrubby grasslands of Africa and Australia.
Colony A group of animals that live together, sharing some aspect of their lives.
Den The name given to the home of a bear.
Dray The name for a squirrel's nest.
Earth The name for a fox's winter home, which is usually a disused rabbit warren.
Eaves The edge of a roof that stands out beyond the wall of a house.
Hibernate To spend the winter asleep.
Larva (plural larvae) The young stage of an insect or frog after it has hatched from an egg.
Litter The young produced at any one time by a mammal.
Paralyzed Unable to move.
Parasite An animal or plant that lives by taking the food or the blood of another animal, without immediately killing it.
Predator An animal that hunts and eats other animals.
Saliva A liquid produced in the mouth to help digest food. Several different types of animals use saliva as glue for building their homes.
Set The name for a badger's home.
Squatters People who move into a house that does not belong to them.
Ventilation shaft A passageway in a building that brings in fresh air from outside.
Warren A series of underground linked tunnels lived in by animals such as rabbits and marmots.
Worker A non-breeding adult female in a colony of ants, termites, wasps or bees.

Further information

The following books will tell you more about animal homes:

Animal Homes. Published by National Geographic, 1981.
Animal Homemakers by Aurelius Battaglia. Putnam Publishing Group, 1978.
The Animal Shelter by Patricia Curtis. Lodestar Books, 1984.
Animal Territories by Daniel Cohen. Hastings House, 1975.
Animals of the Fields and Meadows by Julie Becker. E M C Publishers, 1977.
Animals of the Woods and Forests by Julie Becker. E M C Publishers, 1977.
The Answer Book about Animals by Mary Elting. Putnam Publishing Group, 1984.
How Animals Live by Civardi and Kilpatrick. E D C Publishers, 1981.
The Secret World of Underground Creatures by Dorothy Leon. Julian Messner, 1982.

To find out more about how animals behave in the wild, look out for the interesting wildlife films often shown on television. There are also wildlife parks you can visit.

You may want to join an organization that helps to protect wild animals. Some useful addresses are:

Audubon Naturalist Society of the Central Atlantic States
8940 Jones Mill Road
Chevy Chase, Maryland 20815
301–652–9188

The Conservation Foundation
1717 Massachusetts Avenue, N.W.
Washington, D.C. 20036
202–797–4300

Greenpeace (Canada)
2623 West 4th Avenue
Vancouver, B.C.
V6K 1P9

Greenpeace (USA)
1611 Connecticut Avenue, N.W.
Washington, D.C. 20009
202–462–1177

The Humane Society of the USA
2100 L Street, N.W.
Washington, D.C. 20037
202–452–1100

The International Fund for Animal Welfare
P.O. Box 193
Yarmouth Port, Massachusetts 02675
617–362–4944

National Wildlife Federation
1412 16th Street, N.W.
Washington, D.C. 20036
202–797–6800

Picture acknowledgments

The publishers would like to thank all those who provided photographs on the following pages: Survival Anglia Limited 5 (Stone/Deeble), 11 (Caroline Brett), 12 (Dieter and Mary Plage) and 17 (Jen and Des Bartlett).

Index

Acacia rats 26
Aeries 18
Africa 5, 17, 18
Antelope 5
Ants 5, 10, 20
  leaf-cutter 7
  tailor 20
Arctic 4, 7
Asia 15, 21
Australia 7, 15, 20

Badgers 7, 15, 26
Barbets 22
Bats 29
Bears 7
  polar 7
Beavers 24, 29
Bees 6
Beetles
  carpet 28
  death-watch 28
  dung 6
  tiger 6
  woodworm 28
Birds 4, 6, 9, 16, 17, 18, 19, 21, 22–23, 24, 26
Breeding season 4, 10, 19
Burrows 6, 7, 13, 14, 15, 22, 24

Central America 9
Chimpanzees 19
China 21
Crows 18
Cuckoos 26

Dens 7
Drays 19
Duckbilled platypus 15

Eagles 18
Earth 26
Eggs 6, 9, 10, 13, 15, 23, 24, 25, 26, 29
Egyptian geese 26
Europe 15

Fiddler crabs 13
Fish 5, 24
  blenny 5
  goby 24
Foxes 26

Gorillas 19
Grass, homes of 4, 5, 16–17, 18
Grebes 24
Gray kestrels 26

Hermit crabs 5
House martins 9, 29

India 21
Inuit 4

Jackdaws 18

Leaves, homes of 20–21
Lodges 24

Magpies 18
Marmots 15
Mice 6
  harvest 17
  house 28
Moles 7
Mongooses 26
Moths, clothes 28, 29
Mud, homes of 4, 8, 9, 10, 13

Nests 4, 6, 8, 9, 14, 16, 17, 19, 20, 21, 22, 24, 26
North America 9, 15

Orangutans 19
Ovenbirds 9

Parasites 26
  fleas 26
  lice 26
Parrots 26
Puffins 6

Rabbits 5, 14, 26

Sand, homes of 12–13
Sand martins 6
Sand-mason worms 12
Sets 15, 26
Sleeping nests 19
South America 7, 9
Spiders
  house 28
  leaf-curling 20, 21
  trapdoor 13
  water 24
Squirrels 19
Starlings 26
Sticks, homes of 4, 5, 18–19, 26
Storks, hammer-headed 18, 26
Swallows 9

Tailorbirds 21
Termites 10, 11, 26
Tree holes 10, 22–23, 26

Voles 6

Warblers 16
Warrens 5, 14, 15, 26
Wasps 6, 8, 9, 13, 23, 26
  potter 8
Weaverbirds 17, 29
Whales 5
Wombats 7
Woodpeckers 22, 23, 26